Grandmother thinks
plums are something grand.

"Plums have endless
goodness!" she sings. "Plums
have a lovely sweetness!"

Grandmother sets the plums upon a plate.

"Look! Are they not fine?" she asks.

“Perhaps,” I answer her softly. “But I myself like apples!”

Grandmother picks up a plum. The plum is smooth and spotless. “Don’t you want to try this?” she asks.

I answer shyly, “I don’t think I like plums.”

“That cannot be!” she says sadly.

"Did you not have any plum pancakes? I myself love pancakes with plum jam!"

"Will you try something?" Grandmother asks. "Will you try a plum milkshake? I will gladly make it myself!"

"Plum milkshake?" I look at her oddly.

"Tiptop to the last drop!" she answers. Grandmother blends a plum milkshake in the blender.

I sip the milkshake as Grandmother smiles widely. Grandmother is right! Plums are something grand!

The End